PENGUIN BOOKS — GREAT IDEAS

Why Vegan?

T0200289

Peter Singer

Born 1946

Peter Singer

Why Vegan?

PENGUIN BOOKS — GREAT IDEAS

PENGUIN BOOKS

UK | USA | Canada | Ireland | Australia
India | New Zealand | South Africa

Penguin Books is part of the Penguin Random House group
of companies whose addresses can be found at
global.penguinrandomhouse.com.

Essay 1 is taken from *Animal Liberation* by Peter Singer, published by Bodley
Head, reprinted by permission of The Random House Group Limited. © 2015;
Essay 2 was first published in the *New York Review of Books* (April 5, 1973); Essay 3
(co-written with Jim Mason) is taken from *Eating: What we eat and why it matters*
by Jim Mason and Peter Singer, published by Arrow, reprinted by permission of
The Random House Group Limited. © 2006; Essay 4 was published in the
International Journal for the Study of Animal Problems 3(1), (1982); Essay 5 was first
published in *Consuming Passions* (1998), edited by Sian Griffiths and Jennifer
Wallace; Essay 6 was first published by Project Syndicate (2010); Essay 7 was first
published in *Free Inquiry,* a publication of the Council for Secular Humanism, a
program of the Center for Inquiry, under the title 'A Case for Veganism' (2007);
Essay 8 was first published in *Leapsmag* online (2018). The final essay (co-written
with Paola Cavalieri) was first published by Project Syndicate (2020).

This selection published in Penguin Books 2020

001

Set in 11.2/13.75 pt Dante MT Std
Typeset by Jouve (UK), Milton Keynes
Printed and bound in Great Britain by Clays Ltd, Elcograf S.p.A.

A CIP catalogue record for this book
is available from the British Library

ISBN: 978-0-241-47238-5

www.greenpenguin.co.uk

Contents

Animal Liberation: Preface to the 1975 Edition

This book is about the tyranny of human over non-human animals. This tyranny has caused and today is still causing an amount of pain and suffering that can only be compared with that which resulted from the centuries of tyranny by white humans over black humans. The struggle against this tyranny is a struggle as important as any of the moral and social issues that have been fought over in recent years.

Most readers will take what they have just read to be a wild exaggeration. Five years ago I myself would have laughed at the statements I have now written in complete seriousness. Five years ago I did not know what I know today. If you read this book carefully, paying special attention to the second and third chapters, you will then know as much of what I know about the oppression of animals as it is possible to get into a book of reasonable length. Then you will be able to judge if my opening paragraph is a wild exaggeration or a sober estimate of a situation largely unknown to the general public. So I do not ask you to believe my opening paragraph now. All I ask is that you reserve your judgment until you have read the book.

*

Soon after I began work on this book my wife and I were invited to tea – we were living in England at the time – by a lady who had heard that I was planning to write about animals. She herself was very interested in animals, she said, and she had a friend who had already written a book about animals and would be *so* keen to meet us.

When we arrived our hostess's friend was already there, and she certainly was keen to talk about animals. 'I do love animals,' she began. 'I have a dog and two cats, and do you know they get on together wonderfully well. Do you know Mrs Scott? She runs a little hospital for sick pets . . .' and she was off. She paused while refreshments were served, took a ham sandwich, and then asked us what pets we had.

We told her we didn't own any pets. She looked a little surprised, and took a bite of her sandwich. Our hostess, who had now finished serving the sandwiches, joined us and took up the conversation: 'But you *are* interested in animals, aren't you, Mr Singer?'

We tried to explain that we were interested in the prevention of suffering and misery; that we were opposed to arbitrary discrimination; that we thought it wrong to inflict needless suffering on another being, even if that being were not a member of our own species; and that we believed animals were ruthlessly and cruelly exploited by humans, and we wanted this changed. Otherwise, we said, we were not especially 'interested in' animals. Neither of us had ever been inordinately fond of dogs, cats, or horses in the way that many people are. We didn't 'love' animals. We simply wanted them treated as the independent sentient beings that they are, and not as a means to human ends – as the pig whose flesh was now in our hostess's sandwiches had been treated.

This book is not about pets. It is not likely to be comfortable reading for those who think that love for animals involves no more than stroking a cat or feeding birds in the garden. It is intended rather for people who are concerned about ending oppression and exploitation wherever they occur, and in seeing that the basic moral principle of equal consideration of interests is not arbitrarily restricted to members of our own species. The assumption that in order to be interested in such matters one must be an 'animal-lover' is itself an indication of the absence of the slightest inkling that the moral standards that we apply among human beings might extend to other animals. No one, except a racist concerned to smear his opponents, would suggest that in order to be concerned about equality for mistreated racial minorities you have to love those minorities, or regard them as cute and cuddly. So why make this assumption about people who work for improvements in the conditions of animals?

The portrayal of those who protest against cruelty to animals as sentimental, emotional 'animal-lovers' has had the effect of excluding the entire issue of our treatment of nonhumans from serious political and moral discussion. It is easy to see why we do this. If we did give the issue serious consideration, if, for instance, we looked closely at the conditions in which animals live in the modern 'factory farms' that produce our meat, we might be made uncomfortable about ham sandwiches, roast beef, fried chicken, and all those other items in our diet that we prefer not to think of as dead animals.

This book makes no sentimental appeals for sympathy toward 'cute' animals. I am no more outraged by the slaughter of horses or dogs for meat than I am by the slaughter of pigs for this purpose. When the United States Defense Department finds that its use of beagles to test lethal gases has evoked a howl of protest and offers to use rats instead, I am not appeased.

This book is an attempt to think through, carefully and consistently, the question of how we ought to treat nonhuman animals. In the process it exposes the prejudices that lie behind our present attitudes and behavior. In the chapters that describe what these attitudes mean in practical terms – how animals suffer from the tyranny of human beings – there are passages that will arouse some emotions. These will, I hope, be emotions of anger and outrage, coupled with a determination to do something about the practices described. Nowhere in this book, however, do I appeal to the reader's emotions where they cannot be supported by reason. When there are unpleasant things to be described it would be dishonest to try to describe them in some neutral way that hid their real unpleasantness. You cannot write objectively about the experiments of the Nazi concentration camp 'doctors' on those they considered 'subhuman' without stirring emotions; and the same is true of a description of some of the experiments performed today on nonhumans in laboratories in America, Britain, and elsewhere. The ultimate justification for opposition to both these kinds of experiments, though, is not emotional. It is an appeal to basic moral principles which we all accept, and the application of these principles to the victims of both kinds of experiment is demanded by reason, not emotion.

The title of this book has a serious point behind it. A liberation movement is a demand for an end to prejudice and discrimination based on an arbitrary characteristic like race or sex. The classic instance is the Black Liberation movement. The immediate appeal of this movement, and its initial, if limited, success, made it a model for other oppressed groups. We soon became familiar with Gay Liberation and movements on behalf of American Indians and Spanish-speaking Americans. When a majority group – women – began their campaign some thought we had come to the end of the road. Discrimination on the basis of sex, it was said, was the last form of discrimination to be universally accepted and practiced without secrecy or pretense, even in those liberal circles that have long prided themselves on their freedom from prejudice against racial minorities.

We should always be wary of talking of 'the last remaining form of discrimination.' If we have learned anything from the liberation movements we should have learned how difficult it is to be aware of latent prejudices in our attitudes to particular groups until these prejudices are forcefully pointed out to us.

A liberation movement demands an expansion of our moral horizons. Practices that were previously regarded as natural and inevitable come to be seen as the result of an unjustifiable prejudice. Who can say with any confidence that none of his or her attitudes and practices can legitimately be questioned? If we wish to avoid being numbered among the oppressors, we must be prepared to rethink all our attitudes to other groups, including the most fundamental of them. We need to consider our attitudes from the point of view of those who suffer by them, and by the practices that follow from them. If we can make this unaccustomed mental switch we may discover a pattern in our attitudes and practices that operates so as consistently to benefit the same group – usually the group to which we ourselves belong – at the expense of another group. So we come to see that there is a case for a new liberation movement.

The aim of this book is to lead you to make this mental switch in your attitudes and practices toward a very large group of beings: members of species other than our own. I believe that our present attitudes to these beings are based on a long history of prejudice and arbitrary discrimination. I argue that there can be no reason – except the selfish desire to preserve the privileges of the exploiting group – for refusing to extend the basic principle of equality of consideration to members of other species. I ask you to recognize that your attitudes to members of other species are a form of prejudice no less objectionable than prejudice about a person's race or sex.

In comparison with other liberation movements, Animal Liberation has a lot of handicaps. First and most obvious is the fact that members of the exploited group cannot themselves make an organized protest against the treatment they receive (though they can and do protest to the best of their abilities individually). We have to speak up on behalf of those who cannot speak for themselves. You can appreciate how serious this handicap is by asking yourself how long blacks would have had to wait for equal rights if they had not been able to stand up for themselves and demand it. The less able a group is to stand up and organize against oppression, the more easily it is oppressed.

More significant still for the prospects of the Animal Liberation movement is the fact that almost all of the oppressing group are directly involved in, and see themselves as benefiting from, the oppression. There are few humans indeed who can view the oppression of animals with the detachment possessed, say, by Northern whites debating the institution of slavery in the Southern states of the Union. People who eat pieces of slaughtered nonhumans every day find it hard to believe that they are doing wrong; and they also find it hard to imagine what else they could eat. On this issue, anyone who eats meat is an interested party. They benefit – or at least they think they benefit – from the present disregard of the interests of nonhuman animals. This makes persuasion more difficult. How many Southern slaveholders were

persuaded by the arguments used by the Northern abo-
litionists, and accepted by nearly all of us today? Some,
but not many. I can and do ask you to put aside your
interest in eating meat when considering the arguments
of this book; but I know from my own experience that
with the best will in the world this is not an easy thing
to do. For behind the mere momentary desire to eat
meat on a particular occasion lie many years of habitual
meat-eating which have conditioned our attitudes to
animals.

Habit. That is the final barrier that the Animal Liberation movement faces. Habits not only of diet but also
of thought and language must be challenged and
altered. Habits of thought lead us to brush aside descriptions of cruelty to animals as emotional, for 'animal-lovers
only'; or if not that, then anyway the problem is so trivial
in comparison to the problems of human beings that no
sensible person could give it time and attention. This
too is a prejudice – for how can one know that a problem
is trivial until one has taken the time to examine its
extent? Although in order to allow a more thorough
treatment this book deals with only two of the many
areas in which humans cause other animals to suffer, I
do not think anyone who reads it to the end will ever
again think that the only problems that merit time and
energy are problems concerning humans.

The habits of thought that lead us to disregard the
interests of animals can be challenged, as they are challenged in the following pages. This challenge has to be
expressed in a language, which in this case happens to

be English. The English language, like other languages, reflects the prejudices of its users. So authors who wish to challenge these prejudices are in a well-known type of bind: either they use language that reinforces the very prejudices they wish to challenge, or else they fail to communicate with their audience. This book has already been forced along the former of these paths. We commonly use the word 'animal' to mean 'animals other than human beings.' This usage sets humans apart from other animals, implying that we are not ourselves animals – an implication that everyone who has had elementary lessons in biology knows to be false.

In the popular mind the term 'animal' lumps together beings as different as oysters and chimpanzees, while placing a gulf between chimpanzees and humans, although our relationship to those apes is much closer than the oyster's. Since there exists no other short term for the nonhuman animals, I have, in the title of this book and elsewhere in these pages, had to use 'animal' as if it did not include the human animal. This is a regrettable lapse from the standards of revolutionary purity but it seems necessary for effective communication. Occasionally, however, to remind you that this is a matter of convenience only, I shall use longer, more accurate modes of referring to what was once called 'the brute creation.' In other cases, too, I have tried to avoid language which tends to degrade animals or disguise the nature of the food we eat.

The basic principles of Animal Liberation are very simple. I have tried to write a book that is clear and easy to understand, requiring no expertise of any kind. It is necessary, however, to begin with a discussion of the principles that underlie what I have to say. While there should be nothing here that is difficult, readers unused to this kind of discussion might find the first chapter rather abstract. Don't be put off. In the next chapters we get down to the little-known details of how our species oppresses others under our control. There is nothing abstract about this oppression, or about the chapters that describe it.

If the recommendations made in the following chapters are accepted, millions of animals will be spared considerable pain. Moreover, millions of humans will benefit too. As I write, people are starving to death in many parts of the world; and many more are in imminent danger of starvation. The United States government has said that because of poor harvests and diminished stocks of grain it can provide only limited – and inadequate – assistance; but as Chapter 4 of this book makes clear, the heavy emphasis in affluent nations on rearing animals for food wastes several times as much food as it produces. By ceasing to rear and kill animals for food, we can make so much extra food available for humans that, properly distributed, it would eliminate starvation and malnutrition from this planet. Animal Liberation is Human Liberation too.

Animal Liberation

Animals, Men and Morals is a manifesto for an Animal Liberation movement. The contributors to the book may not all see the issue this way. They are a varied group. Philosophers, ranging from professors to graduate students, make up the largest contingent. There are five of them, including the three editors, and there is also an extract from the unjustly neglected German philosopher with an English name, Leonard Nelson, who died in 1927. There are essays by two novelist/critics, Brigid Brophy and Maureen Duffy, and another by Muriel the Lady Dowding, widow of Dowding of Battle of Britain fame, and the founder of 'Beauty without Cruelty,' a movement that campaigns against the use of animals for furs and cosmetics. The other pieces are by a psychologist, a botanist, a sociologist, and Ruth Harrison, who is probably best described as a professional campaigner for animal welfare.

Whether or not these people, as individuals, would all agree that they are launching a liberation movement for animals, the book as a whole amounts to no less. It is a demand for a complete change in our attitudes to nonhumans. It is a demand that we cease to regard the exploitation of other species as natural and inevitable,

and that, instead, we see it as a continuing moral outrage. Patrick Corbett, Professor of Philosophy at Sussex
University, captures the spirit of the book in his closing
words:

> . . . we require now to extend the great principles
> of liberty, equality and fraternity over the lives
> of animals. Let animal slavery join human slav-
> ery in the graveyard of the past.

The reader is likely to be skeptical. 'Animal Liberation'
sounds more like a parody of liberation movements
than a serious objective. The reader may think: We support the claims of blacks and women for equality
because blacks and women really are equal to whites
and males – equal in intelligence and in abilities, capacity for leadership, rationality, and so on. Humans and
nonhumans obviously are not equal in these respects.
Since justice demands only that we treat equals equally,
unequal treatment of humans and nonhumans cannot
be an injustice.

This is a tempting reply, but a dangerous one. It com-
mits the non-racist and non-sexist to a dogmatic belief
that blacks and women really are just as intelligent, able,
etc., as whites and males – and no more. Quite possibly
this happens to be the case. Certainly attempts to prove
that racial or sexual differences in these respects have a
genetic origin have not been conclusive. But do we
really want to stake our demand for equality on the
assumption that there are no genetic differences of this
kind between the different races or sexes? Surely the

appropriate response to those who claim to have found evidence for such genetic differences is not to stick to the belief that there are no differences, whatever the evidence to the contrary; rather one should be clear that the claim to equality does not depend on IQ. Moral equality is distinct from factual equality. Otherwise it would be nonsense to talk of the equality of human beings, since humans, as individuals, obviously differ in intelligence and almost any ability one cares to name. If possessing greater intelligence does not entitle one human to exploit another, why should it entitle humans to exploit nonhumans?

Jeremy Bentham expressed the essential basis of equality in his famous formula: 'Each to count for one and none for more than one.' In other words, the interests of every being that has interests are to be taken into account and treated equally with the like interests of any other being. Other moral philosophers, before and after Bentham, have made the same point in different ways. Our concern for others must not depend on whether they possess certain characteristics, though just what that concern involves may, of course, vary according to such characteristics.

Bentham, incidentally, was well aware that the logic of the demand for racial equality did not stop at the equality of humans. He wrote:

> The day *may* come when the rest of the animal creation may acquire those rights which never could have been withholden from them but by

the hand of tyranny. The French have already discovered that the blackness of the skin is no reason why a human being should be abandoned without redress to the caprice of a tormentor. It may one day come to be recognized that the number of the legs, the villosity of the skin, or the termination of the *os sacrum*, are reasons equally insufficient for abandoning a sensitive being to the same fate. What else is it that should trace the insuperable line? Is it the faculty of reason, or perhaps the faculty of discourse? But a full-grown horse or dog is beyond comparison a more rational, as well as a more conversable animal, than an infant of a day, or a week, or even a month, old. But suppose they were otherwise, what would it avail? The question is not, Can they *reason*? nor Can they *talk*? but, Can they *suffer*?[1]

Surely Bentham was right. If a being suffers, there can be no moral justification for refusing to take that suffering into consideration, and, indeed, to count it equally with the like suffering (if rough comparisons can be made) of any other being.

So the only question is: do animals other than man suffer? Most people agree unhesitatingly that animals like cats and dogs can and do suffer, and this seems also

[1] *The Principles of Morals and Legislation*, Ch. XVII, Sec. 1, footnote to paragraph 4.

to be assumed by those laws that prohibit wanton cruelty to such animals. Personally, I have no doubt at all about this and find it hard to take seriously the doubts that a few people apparently do have. The editors and contributors of *Animals, Men and Morals* seem to feel the same way, for although the question is raised more than once, doubts are quickly dismissed each time. Nevertheless, because this is such a fundamental point, it is worth asking what grounds we have for attributing suffering to other animals.

It is best to begin by asking what grounds any individual human has for supposing that other humans feel pain. Since pain is a state of consciousness, a 'mental event,' it can never be directly observed. No observations, whether behavioral signs such as writhing or screaming or physiological or neurological recordings, are observations of pain itself. Pain is something one feels, and one can only infer that others are feeling it from various external indications. The fact that only philosophers are ever skeptical about whether other humans feel pain shows that we regard such inference as justifiable in the case of humans.

Is there any reason why the same inference should be unjustifiable for other animals? Nearly all the external signs which lead us to infer pain in other humans can be seen in other species, especially 'higher' animals such as mammals and birds. Behavioral signs – writhing, yelping, or other forms of calling, attempts to avoid the source of pain, and many others – are present. We know, too, that these animals are biologically similar in the relevant respects, having nervous systems like ours which can be observed to function as ours do.

So the grounds for inferring that these animals can feel pain are nearly as good as the grounds for inferring other humans do. Only nearly, for there is one behavioral sign that humans have but nonhumans, with the exception of one or two specially raised chimpanzees, do not have. This, of course, is a developed language. As the quotation from Bentham indicates, this has long been regarded as an important distinction between man and other animals. Other animals may communicate with each other, but not in the way we do. Following Chomsky, many people now mark this distinction by saying that only humans communicate in a form that is governed by rules of syntax. (For the purposes of this argument, linguists allow those chimpanzees who have learned a syntactic sign language to rank as honorary humans.) Nevertheless, as Bentham pointed out, this distinction is not relevant to the question of how animals ought to be treated, unless it can be linked to the issue of whether animals suffer.

This link may be attempted in two ways. First, there is a hazy line of philosophical thought, stemming perhaps from some doctrines associated with Wittgenstein, which maintains that we cannot meaningfully attribute states of consciousness to beings without language. I have not seen this argument made explicit in print, though I have come across it in conversation. This position seems to me very implausible, and I doubt that it would be held at all if it were not thought to be a consequence of a broader view of the significance of language. It may be that the use of a public, rule-governed language is a precondition of conceptual thought. It may even be, although personally I doubt it, that we cannot meaningfully speak of a creature having an intention unless that creature can use a language. But states like pain, surely, are more primitive than either of these, and seem to have nothing to do with language.

Indeed, as Jane Goodall points out in her study of chimpanzees, when it comes to the expression of feelings and emotions, humans tend to fall back on non-linguistic modes of communication which are often found among apes, such as a cheering pat on the back, an exuberant embrace, a clasp of hands, and so on.[2] Michael Peters makes a similar point in his contribution to *Animals, Men and Morals* when he notes that the basic signals we use to convey pain, fear, sexual arousal, and so on are not specific to our species. So there seems to be no reason at all to believe that a creature without language cannot suffer.

The second, and more easily appreciated, way of linking language and the existence of pain is to say that the best evidence that we can have that another creature is in pain is when he tells us that he is. This is a distinct line of argument, for it is not being denied that a non-language-user conceivably could suffer, but only that we

2 Jane van Lawick-Goodall, *In the Shadow of Man* (Houghton Mifflin, 1971), p.225.

could know that he is suffering. Still, this line of argument seems to me to fail, and for reasons similar to those just given. 'I am in pain' is not the best possible evidence that the speaker is in pain (he might be lying) and it is certainly not the only possible evidence. Behavioral signs and knowledge of the animal's biological similarity to ourselves together provide adequate evidence that animals do suffer. After all, we would not accept linguistic evidence if it contradicted the rest of the evidence. If a man was severely burned, and behaved as if he were in pain, writhing, groaning, being very careful not to let his burned skin touch anything, and so on, but later said he had not been in pain at all, we would be more likely to conclude that he was lying or suffering from amnesia than that he had not been in pain.

Even if there were stronger grounds for refusing to attribute pain to those who do not have a language, the consequences of this refusal might lead us to examine these grounds unusually critically. Human infants, as well as some adults, are unable to use language. Are we to deny that a year-old infant can suffer? If not, how can language be crucial? Of course, most parents can understand the responses of even very young infants better than they understand the responses of other animals, and sometimes infant responses can be understood in the light of later development.

This, however, is just a fact about the relative knowledge we have of our own species and other species, and most of this knowledge is simply derived from closer contact. Those who have studied the behavior of other animals soon learn to understand their responses at least as well as we understand those of an infant. (I am not just referring to Jane Goodall's and other well-known studies of apes. Consider, for example, the degree of understanding achieved by Tinbergen from watching herring gulls.)[3] Just as we can understand infant human behavior in the light of adult human behavior, so we can understand the behavior of other species in the light of our own behavior (and sometimes we can understand our own behavior better in the light of the behavior of other species).

The grounds we have for believing that other mammals and birds suffer are, then, closely analogous to the grounds we have for believing that other humans suffer. It remains to consider how far down the evolutionary scale this analogy holds. Obviously it becomes poorer when we get further away from man. To be more precise would require a detailed examination of all that we know about other forms of life. With fish, reptiles, and other vertebrates the analogy still seems strong, with molluscs like oysters it is much weaker. Insects are more difficult, and it may be that in our present state of knowledge we must be agnostic about whether they are capable of suffering.

If there is no moral justification for ignoring suffering when it occurs, and it does occur in other species, what are we to say of our attitudes toward these other

3 N. Tinbergen, *The Herring Gull's World* (Basic Books, 1961).

species? Richard Ryder, one of the contributors to *Animals, Men and Morals*, uses the term 'speciesism' to describe the belief that we are entitled to treat members of other species in a way in which it would be wrong to treat members of our own species. The term is not euphonious, but it neatly makes the analogy with racism. The non-racist would do well to bear the analogy in mind when he is inclined to defend human behavior toward nonhumans. 'Shouldn't we worry about improving the lot of our own species before we concern ourselves with other species?' he may ask. If we substitute 'race' for 'species' we shall see that the question is better not asked. 'Is a vegetarian diet nutritionally adequate?' resembles the slave-owner's claim that he and the whole economy of the South would be ruined without slave labor. There is even a parallel with skeptical doubts about whether animals suffer, for some defenders of slavery professed to doubt whether blacks really suffer in the way that whites do.

I do not want to give the impression, however, that the case for Animal Liberation is based on the analogy with racism and no more. On the contrary, *Animals, Men and Morals* describes the various ways in which humans exploit nonhumans, and several contributors consider the defenses that have been offered, including the defense of meat-eating mentioned in the last paragraph. Sometimes the rebuttals are scornfully dismissive, rather than carefully designed to convince the detached critic. This may be a fault, but it is a fault that is inevitable, given the kind of book this is. The issue is not one on which one can remain detached. As the editors state in their Introduction:

> Once the full force of moral assessment has been made explicit there can be no rational excuse left for killing animals, be they killed for food, science, or sheer personal indulgence. We have not assembled this book to provide the reader with yet another manual on how to make brutalities less brutal. Compromise, in the traditional sense of the term, is simple unthinking weakness when one considers the actual reasons for our crude relationships with the other animals.

The point is that on this issue there are few critics who are genuinely detached. People who eat pieces of slaughtered nonhumans every day find it hard to believe that they are doing wrong; and they also find it hard to imagine what else they could eat. So for those who do not place nonhumans beyond the pale of morality, there comes a stage when further argument seems pointless, a stage at which one can only accuse one's opponent of hypocrisy and reach for the sort of sociological account of our practices and the way we defend them that is attempted by David Wood in his contribution to this book. On the other hand, to those unconvinced by the arguments, and unable to accept that they are merely rationalizing their dietary preferences and their fear of being thought peculiar, such sociological explanations can only seem insultingly arrogant.

II

The logic of speciesism is most apparent in the practice of experimenting on nonhumans in order to benefit humans. This is because the issue is rarely obscured by allegations that nonhumans are so different from humans that we cannot know anything about whether they suffer. The defender of vivisection cannot use this argument because he needs to stress the similarities between man and other animals in order to justify the usefulness to the former of experiments on the latter. The researcher who makes rats choose between starvation and electric shocks to see if they develop ulcers (they do) does so because he knows that the rat has a nervous system very similar to man's, and presumably feels an electric shock in a similar way.

Richard Ryder's restrained account of experiments on animals made me angrier with my fellow men than anything else in this book. Ryder, a clinical psychologist by profession, himself experimented on animals before he came to hold the view he puts forward in his essay. Experimenting on animals is now a large industry, both academic and commercial. In 1969, more than 5 million experiments were performed in Britain, the vast majority without anesthetic (though how many of these involved pain is not known). There are no accurate US figures, since there is no federal law on the subject, and in many cases no state law either. Estimates vary from 20 million to 200 million. Ryder suggests that 80 million may be the best guess. We tend to think that this is all for vital medical research, but of course it is not. Huge numbers of animals are used in university departments from Forestry to Psychology, and even more are used for commercial purposes, to test whether cosmetics can cause skin damage, or shampoos eye damage, or to test food additives or laxatives or sleeping pills or anything else.

A standard test for foodstuffs is the 'LD50.' The object of this test is to find the dosage level at which 50 percent of the test animals will die. This means that nearly all of them will become very sick before finally succumbing or surviving. When the substance is a harmless one, it may be necessary to force huge doses down the animals, until in some cases sheer volume or concentration causes death.

Ryder gives a selection of experiments, taken from recent scientific journals. I will quote two, not for the sake of indulging in gory details, but in order to give an idea of what normal researchers think they may legitimately do to other species. The point is not that the individual researchers are cruel men, but that they are behaving in a way that is allowed by our speciesist attitudes. As Ryder points out, even if only 1 percent of the experiments involve severe pain, that is 50,000 experiments in Britain each year, or nearly 150 every day (and about fifteen times as many in the United States, if Ryder's guess is right). Here then are two experiments:

O. S. Ray and R. J. Barrett of Pittsburg gave electric

shocks to the feet of 1,042 mice. They then caused con-
vulsions by giving more intense shocks through
cup-shaped electrodes applied to the animals' eyes or
through pressure spring clips attached to their ears.
Unfortunately some of the mice who 'successfully com-
pleted Day One training were found sick or dead prior
to testing on Day Two.' (*Journal of Comparative and Physiological Psychology*, 1969, vol. 67, pp. 110–116)

At the National Institute for Medical Research, Mill
Hill, London, W. Feldberg and S. L. Sherwood injected
chemicals into the brains of cats – 'with a number of
widely different substances, recurrent patterns of reac-
tion were obtained. Retching, vomiting, defaecation,
increased salivation and greatly accelerated respiration
leading to panting were common features.' . . .

The injection into the brain of a large dose of Tubo-
curaine caused the cat to jump 'from the table to the
floor and then straight into its cage, where it started
calling more and more noisily whilst moving about rest-
lessly and jerkily . . . finally the cat fell with legs and
neck flexed, jerking in rapid clonic movements, the con-
dition being that of a major [epileptic] convulsion . . .
within a few seconds the cat got up, ran for a few yards
at high speed and fell in another fit. The whole process
was repeated several times within the next ten minutes,
during which the cat lost faeces and foamed at the
mouth.'

This animal finally died thirty-five minutes after the
brain injection. (*Journal of Physiology*, 1954, vol. 123,
pp.148–167)

There is nothing secret about these experiments. One has only to open any recent volume of a learned journal, such as the *Journal of Comparative and Physiological Psychology*, to find full descriptions of experiments of this sort, together with the results obtained – results that are frequently trivial and obvious. The experiments are often supported by public funds.

It is a significant indication of the level of acceptability of these practices that, although these experiments are taking place at this moment on university campuses throughout the country, there has, so far as I know, not been the slightest protest from the student movement. Students have been rightly concerned that their universities should not discriminate on grounds of race or sex, and that they should not serve the purposes of the military or big business. Speciesism continues undisturbed, and many students participate in it. There may be a few qualms at first, but since everyone regards it as normal, and it may even be a required part of a course, the student soon becomes hardened and, dismissing his earlier feelings as 'mere sentiment,' comes to regard animals as statistics rather than sentient beings with interests that warrant consideration.

Argument about vivisection has often missed the point because it has been put in absolutist terms: would the abolitionist be prepared to let thousands die if they could be saved by experimenting on a single animal? The way to reply to this purely hypothetical question is to pose another: Would the experimenter be prepared to experiment on a human orphan under six months old, if it were the only way to save many lives? (I say 'orphan' to avoid the complication of parental feelings, although in doing so I am being overfair to the experimenter, since the nonhuman subjects of experiments are not orphans.) A negative answer to this question indicates that the experimenter's readiness to use nonhumans is simple discrimination, for adult apes, cats, mice, and other mammals are more conscious of what is happening to them, more self-directing, and, so far as we can tell, just as sensitive to pain as a human infant. There is no characteristic that human infants possess that adult mammals do not have to the same or a higher degree.

(It might be possible to hold that what makes it wrong to experiment on a human infant is that the infant will in time develop into more than the nonhuman, but one would then, to be consistent, have to oppose abortion, and perhaps contraception, too, for the fetus and the egg and sperm have the same potential as the infant. Moreover, one would still have no reason for experimenting on a nonhuman rather than a human with brain damage severe enough to make it impossible for him to rise above infant level.)

The experimenter, then, shows a bias for his own species whenever he carries out an experiment on a nonhuman for a purpose that he would not think justified him in using a human being at an equal or lower level of sentience, awareness, ability to be self-directing, etc. No one familiar with the kind of results yielded by these experiments can have the slightest doubt that if this bias were eliminated the number of experiments performed would be zero or very close to it.

III

If it is vivisection that shows the logic of speciesism most clearly, it is the use of other species for food that is at the heart of our attitudes toward them. Most of *Animals, Men and Morals* is an attack on meat-eating – an attack which is based solely on concern for nonhumans, without reference to arguments derived from considerations of ecology, macrobiotics, health, or religion.

The idea that nonhumans are utilities, means to our ends, pervades our thought. Even conservationists who are concerned about the slaughter of wild fowl but not about the vastly greater slaughter of chickens for our tables are thinking in this way – they are worried about what we would lose if there were less wildlife. Stanley Godlovitch, pursuing the Marxist idea that our thinking is formed by the activities we undertake in satisfying our needs, suggests that man's first classification of his environment was into Edibles and Inedibles. Most animals came into the first category, and there they have remained.

Man may always have killed other species for food, but he has never exploited them so ruthlessly as he does today. Farming has succumbed to business methods, the objective being to get the highest possible ratio of output (meat, eggs, milk) to input (fodder, labor costs, etc.). Ruth Harrison's essay 'On Factory Farming' gives an account of some aspects of modern methods, and of the unsuccessful British campaign for effective controls, a campaign which was sparked off by her *Animal Machines* (Stuart: London, 1964).

Her article is in no way a substitute for her earlier book. This is a pity since, as she says, 'Farm produce is still associated with mental pictures of animals browsing in the fields, . . . of hens having a last forage before going to roost . . .' Yet neither in her article nor elsewhere in *Animals, Men and Morals* is this false image replaced by a clear idea of the nature and extent of factory farming. We learn of this only indirectly, when we hear of the code of reform proposed by an advisory committee set up by the British government. Among the proposals, which the government refused to implement on the grounds that they were too idealistic, were: *'Any animal should at least have room to turn around freely.'*

Factory-farm animals need liberation in the most literal sense. Veal calves are kept in stalls five feet by two feet. They are usually slaughtered when about four months old, and have been too big to turn in their stalls for at least a month. Intensive beef herds, kept in stalls only proportionately larger for much longer periods, account for a growing percentage of beef production. Sows are often similarly confined when pregnant, which, because of artificial methods of increasing fertility, can be most of the time. Animals confined in this way do not waste food by exercising, nor do they develop unpalatable muscle.

'A dry bedded area should be provided for all stock.'

Intensively kept animals usually have to stand and sleep on slatted floors without straw, because this makes cleaning easier.

'*Palatable roughage must be readily available to all calves after one week of age.*' In order to produce the pale veal housewives are said to prefer, calves are fed on an all-liquid diet until slaughter, even though they are long past the age at which they would normally eat grass. They develop a craving for roughage, evidenced by attempts to gnaw wood from their stalls. (For the same reason, their diet is deficient in iron.)

'*Battery cages for poultry should be large enough for a bird to be able to stretch one wing at a time.*' Under current British practice, a cage for four or five laying hens has a floor area of twenty inches by eighteen inches, scarcely larger than a double page of the *New York Review of Books*. In this space, on a sloping wire floor (sloping so the eggs roll down, wire so the dung drops through) the birds live for a year or eighteen months while artificial lighting and temperature conditions combine with drugs in their food to squeeze the maximum number of eggs out of them. Table birds are also sometimes kept in cages. More often they are reared in sheds, no less crowded. Under these conditions all the birds' natural activities are frustrated, and they develop 'vices' such as pecking each other to death. To prevent this, beaks are often cut off, and the sheds kept dark.

How many of those who support factory farming by buying its produce know anything about the way it is produced? How many have heard something about it, but are reluctant to check up for fear that it will make them uncomfortable? To non-speciesists, the typical consumer's mixture of ignorance, reluctance to find out the truth, and vague belief that nothing really bad could be allowed seems analogous to the attitudes of 'decent Germans' to the death camps.

There are, of course, some defenders of factory farming. Their arguments are considered, though again rather sketchily, by John Harris. Among the most common: 'Since they have never known anything else, they don't suffer.' This argument will not be put by anyone who knows anything about animal behavior, since he will know that not all behavior has to be learned. Chickens attempt to stretch wings, walk around, scratch, and even dust-bathe or build a nest, even though they have never lived under conditions that allowed these activities. Calves can suffer from maternal deprivation no matter at what age they were taken from their mothers. 'We need these intensive methods to provide protein for a growing population.' As ecologists and famine relief organizations know, we can produce far more protein per acre if we grow the right vegetable crop, soy beans for instance, than if we use the land to grow crops to be converted into protein by animals who use nearly 90 percent of the protein themselves, even when unable to exercise.

There will be many readers of this book who will agree that factory farming involves an unjustifiable degree of exploitation of sentient creatures, and yet will want to say that there is nothing wrong with rearing animals for food, provided it is done 'humanely.' These people are saying, in effect, that although we should not cause animals to suffer, there is nothing wrong with killing them.

There are two possible replies to this view. One is to attempt to show that this combination of attitudes is absurd. Roslind Godlovitch takes this course in her essay, which is an examination of some common attitudes to animals. She argues that from the combination of 'animal suffering is to be avoided' and 'there is nothing wrong with killing animals' it follows that all animal life ought to be exterminated (since all sentient creatures will suffer to some degree at some point in their lives). Euthanasia is a contentious issue only because we place some value on living. If we did not, the least amount of suffering would justify it. Accordingly, if we deny that we have a duty to exterminate all animal life, we must concede that we are placing some value on animal life.

This argument seems to me valid, although one could still reply that the value of animal life is to be derived from the pleasures that life can have for them, so that, provided their lives have a balance of pleasure over pain, we are justified in rearing them. But this would imply that we ought to produce animals and let them live as pleasantly as possible, without suffering.

At this point, one can make the second of the two possible replies to the view that rearing and killing animals for food is all right so long as it is done humanely. This second reply is that so long as we think that a nonhuman may be killed simply so that a human can satisfy his taste for meat, we are still thinking of nonhumans as means rather than as ends in themselves. The factory farm is nothing more than the application of technology to this concept. Even traditional methods involve castration, the separation of mothers and their young, the breaking up of herds, branding or ear-punching, and of course transportation to the abattoirs and the final moments of terror when the animal smells blood and senses danger. If we were to try rearing animals so that they lived and died without suffering, we should find that to do so on anything like the scale of today's meat industry would be a sheer impossibility. Meat would become the prerogative of the rich.

I have been able to discuss only some of the contributions to this book, saying nothing about, for instance, the essays on killing for furs and for sport. Nor have I considered all the detailed questions that need to be asked once we start thinking about other species in the radically different way presented by this book. What, for instance, are we to do about genuine conflicts of interest like rats biting slum children? I am not sure of the answer, but the essential point is just that we *do* see this as a conflict of interests, that we recognize that rats have interests too. Then we may begin to think about other ways of resolving the conflict – perhaps by leaving out rat baits that sterilize the rats instead of killing them.

I have not discussed such problems because they are

side issues compared with the exploitation of other species for food and for experimental purposes. On these central matters, I hope that I have said enough to show that this book, despite its flaws, is a challenge to every human to recognize his attitudes to nonhumans as a form of prejudice no less objectionable than racism or sexism. It is a challenge that demands not just a change of attitudes, but a change in our way of life, for it requires us to become vegetarians.

Can a purely moral demand of this kind succeed? The odds are certainly against it. The book holds out no inducements. It does not tell us that we will become healthier, or enjoy life more, if we cease exploiting animals. Animal Liberation will require greater altruism on the part of mankind than any other liberation movement, since animals are incapable of demanding it for themselves, or of protesting against their exploitation by votes, demonstrations, or bombs. Is man capable of such genuine altruism? Who knows? If this book does have a significant effect, however, it will be a vindication of all those who have believed that man has within himself the potential for more than cruelty and selfishness.

An Ethical Way of Treating Chickens?

WITH JIM MASON

Some people have so little empathy for chickens that they don't care how they are treated. To call a person a 'chicken' is to show contempt for his lack of courage, and to call someone a 'birdbrain' is to suggest exceptional stupidity. But chickens can recognize up to 90 other individual chickens and know whether each one of those birds is higher or lower in the pecking order than they are themselves. Researchers have shown that if they get a small amount of food when they immediately peck at a colored button, but a larger amount if they wait 22 seconds, they can learn to wait before pecking.[1] Moreover, after thousands of generations of domestic breeding, chickens still retain the ability to give and to understand distinct alarm calls, depending on whether there is a threat from above, like a hawk, or from the ground, like a raccoon. When scientists play back a recording of an 'aerial' alarm call, chickens respond differently than when they hear a recording of a 'ground' alarm call.[2]

Interesting as these studies are, the point of real ethical significance is not how clever chickens are, but whether they can suffer – and of that there can be no serious doubt. Chickens have nervous systems similar to ours, and when we do things to them that are likely to hurt a sensitive creature, they show behavioral and physiological responses that are like ours. When stressed or bored, chickens show what scientists call 'stereotypical behavior,' or repeated futile movements, like caged animals who pace back and forth. When they have become acquainted with two different habitats and find one preferable to the other, they will work hard to get to the living quarters they prefer. Lame chickens will choose food to which painkillers have been added; the drug evidently relieves the pain they feel and allows them to be more active.[3]

Most people readily agree that we should avoid inflicting unnecessary suffering on animals. Summarizing the recent research on the mental lives of chickens and other farmed animals, Christine Nicol, professor of animal welfare at Bristol University, in England, has said: 'Our challenge is to teach others that every animal we intend to eat or use is a complex individual, and to adjust our farming culture accordingly.'[4] We are about to see how far that farming culture would have to change to achieve this.

Almost all the chickens sold in supermarkets – known in the industry as 'broilers' – are raised in very large sheds. A typical shed measures 490 feet long by 45 feet wide and will hold 30,000 or more chickens. The National Chicken Council, the trade association for the US chicken industry, issues Animal Welfare Guidelines that indicate a stocking density of 96 square inches for a bird of average market weight[5] – that's about the size of a standard sheet of American 8.5 inch x 11 inch typing paper. When the chicks are small, they are not crowded, but as they near market weight, they cover the floor completely – at first glance, it seems as if the shed is carpeted in white. They are unable to move without pushing through other birds, unable to stretch their wings at will, or to get away from more dominant, aggressive birds. The crowding causes stress, because in a more natural situation, chickens will establish a 'pecking order' and make their own space accordingly.

If the producers gave the chickens more space they would gain more weight and be less likely to die, but it isn't the productivity of each bird – let alone the bird's welfare – that determines how they are kept. As one industry manual explains: 'Limiting the floor space gives poorer results on a per bird basis, yet the question has always been and continues to be: What is the least amount of floor space necessary per bird to produce the greatest return on investment.'[6]

In Britain, a judge ruled in 1997 that crowding chickens like this is cruel. The case arose when McDonald's claimed that two British environmental activists, Helen Steel and David Morris, had libeled the company in a leaflet that, among other things, said that McDonald's was responsible for cruelty. Steel and Morris had no money to pay lawyers to defend themselves against the corporate giant so they ran the case themselves, calling experts to give evidence in support of their claims. The

'McLibel' case turned into the longest trial in English legal history. After hearing many experts testify, the judge, Rodger Bell, ruled that, although some other claims Steel and Morris had made were false, the charge of cruelty was true: 'Broiler chickens which are used to produce meat for McDonald's . . . spend the last few days of their lives with very little room to move,' he said. 'The severe restriction of movement of those last few days is cruel and McDonald's are culpably responsible for that cruel practice.'[7]

ENTER THE CHICKEN SHED (WARNING: MAY BE DISTURBING TO SOME READERS)

Enter a typical chicken shed and you will experience a burning feeling in your eyes and your lungs. That's the ammonia – it comes from the birds' droppings, which are simply allowed to pile up on the floor without being cleaned out, not merely during the growing period of each flock, but typically for an entire year, and sometimes for several years.[8] High ammonia levels give the birds chronic respiratory disease, sores on their feet and hocks, and breast blisters. It makes their eyes water, and when it is really bad, many birds go blind.[9] As the birds, bred for extremely rapid growth, get heavier, it hurts them to keep standing up, so they spend much of their time sitting on the excrement-filled litter – hence the breast blisters.

Chickens have been bred over many generations to

produce the maximum amount of meat in the least amount of time. They now grow three times as fast as chickens raised in the 1950s while consuming one-third as much feed.[10] But this relentless pursuit of efficiency has come at a cost: their bone growth is outpaced by the growth of their muscles and fat. One study found that 90 percent of broilers had detectable leg problems, while 26 percent suffered chronic pain as a result of bone disease.[11] Professor John Webster of the University of Bristol's School of Veterinary Science has said: 'Broilers are the only livestock that are in chronic pain for the last 20 percent of their lives. They don't move around, not because they are overstocked, but because it hurts their joints so much.'[12] Sometimes vertebrae snap, causing paralysis. Paralyzed birds or birds whose legs have collapsed cannot get to food or water, and – because the growers don't bother to, or don't have time to, check on individual birds – die of thirst or starvation. Given these and other welfare problems and the vast number of animals involved – nearly 9 billion in the United States – Webster regards industrial chicken production as, 'in both magnitude and severity, the single most severe, systematic example of man's inhumanity to another sentient animal.'[13]

Criticize industrial farming and industry spokespeople are sure to respond that it is in the interests of those who raise animals to keep them healthy and happy so that they will grow well. Commercial chicken-rearing conclusively refutes this claim. Birds who die prematurely may cost the grower money, but it is the total productivity of the shed that matters. G. Tom Tabler, who manages the Applied Broiler Research Unit at the University of Arkansas, and A. M. Mendenhall, of the Department of Poultry Science at the same university, have posed the question: 'Is it more profitable to grow the biggest bird and have increased mortality due to heart attacks, ascites (another illness caused by fast growth), and leg problems, or should birds be grown slower so that birds are smaller, but have fewer heart, lung, and skeletal problems?' Once such a question is asked, as the researchers themselves point out, it takes only 'simple calculations' to draw the conclusion that, depending on the various costs, often 'it is better to get the weight and ignore the mortality.'[14]

Breeding chickens for rapid growth creates a different problem for the breeder birds, the parents of the chickens people eat. The parents have the same genetic characteristics as their offspring – including huge appetites. But the breeder birds must live to maturity and keep on breeding as long as possible. If they were given as much food as their appetites demand, they would grow grotesquely fat and might die before they became sexually mature. If they survived at all, they would be unable to breed. So breeder operators ration the breeder birds to eat 60 to 80 percent less than their appetites would lead them to eat if they could.[15] The National Chicken Council's Animal Welfare Guidelines refer to 'off-feed days;' that is, days on which the hungry birds get no food at all. This is liable to make them drink 'excessive' amounts of water, so the water, too, can be restricted on those days. They compulsively peck the ground, even when there is nothing there, either to relieve the stress, or in the vain hope of finding something to eat. As Mr Justice Bell, who examined this practice in the McLibel case, said: 'My conclusion is that the practice of rearing breeders for appetite, that is to feel especially hungry, and then restricting their feed with the effect of keeping them hungry, is cruel. It is a well-planned device for profit at the expense of suffering of the birds.'

The fast-growing offspring of these breeding birds live for only six weeks. At that age they are caught, put into crates, and trucked to slaughter. A *Washington Post* journalist observed the catchers at work: 'They grab birds by their legs, thrusting them like sacks of laundry into the cages, sometimes applying a shove.' To do their job more quickly, the catchers pick up only one leg of each bird, so that they can hold four or five chickens in each hand. (The National Chicken Council's Animal Welfare Guidelines, eager to avoid curtailing any practice that may be economically advantageous, says 'The maximum number of birds per hand is five.') Dangling from one leg, the frightened birds flap and writhe and often suffer dislocated and broken hips, broken wings, and internal bleeding.[16]

Crammed into cages, the birds then travel to the slaughterhouse, a journey that can take several hours. When their turn to be removed from the crates finally comes, their feet are snapped into metal shackles hanging from a conveyor belt that moves towards the killing room. Speed is the essence, because the slaughterhouse is paid by the number of pounds of chicken that comes out the end. Today a killing line typically moves at 90 birds a minute, and speeds can go as high as 120 birds a minute, or 7,200 an hour. Even the lower rate is twice as fast as the lines moved twenty years ago. At such speeds, even if the handlers wanted to handle the birds gently and with care, they just couldn't.

In the United States, in contrast to other developed nations, the law does not require that chickens (or ducks, or turkeys) be rendered unconscious before they are slaughtered. As the birds move down the killing line, still upside down, their heads are dipped into an electrified water bath, which in the industry is called 'the stunner.' But this is a misnomer. Dr Mohan Raj, a researcher in the Department of Clinical Veterinary Science at the University of Bristol, in England, has recorded the brain activity of chickens after various forms of stunning and reported his results in such publications as *World's Poultry Science Journal*. We asked him: 'Can the American consumer be confident that broilers he or she buys in a supermarket have been properly stunned so that they are unconscious when they have their throats cut?' His answer was clear: 'No. The majority of broilers are likely to be conscious and suffer pain and distress at slaughter under the existing water bath electrical stunning systems.' He went on to explain that the type of electrical current used in the stunning procedure was not adequate to make the birds immediately unconscious. Using a current that would produce immediate loss of consciousness, however, would risk damage to the quality of the meat. Since there is no legal requirement for stunning, the industry won't take that risk. Instead, the inadequate current that is used evidently paralyzes the birds without rendering them unconscious. From the point of view of the slaughterhouse operator, inducing paralysis is as good as inducing unconsciousness, for it stops the birds from thrashing about and makes it easier to cut their throats.

Because of the fast line speed, even the throat-cutting that follows the electrified water bath misses some birds, and they then go alive and conscious into the next stage of the process, a tank of scalding water. It is difficult to get figures on how many birds are, in effect, boiled alive, but documents obtained under the Freedom of Information act indicate that in the United States alone, it could be as many as three million a year.[17] At that rate, 11 chickens would have been scalded to death in the time it takes you to read this page. But the real figure might be much higher. An undercover videotape made at a Tyson slaughterhouse at Heflin, Alabama, shows dozens of birds who have been mutilated by throat-cutting machines that were not working properly. Workers rip the heads off live chickens that have been missed by the cutting blade. Conscious birds go into the scalding tank. A plant worker is recorded as saying that it is acceptable for 40 birds per shift to be missed by the backup killer and scalded alive.[18]

If you found the last few paragraphs unpleasant reading, Virgil Butler, who spent years working for Tyson

Foods in the killing room of a slaughterhouse in Grannis, Arkansas, killing 80,000 chickens a night, mostly for Kentucky Fried Chicken, says that what we have described 'doesn't even come close to the horrors I have seen.' The killing line on which he worked moved so quickly that it was impossible to kill all the chickens before the line moved on. On a good night, he says, about one in every seven of the chickens were alive when they went into the scalding tank. On an average night, it might be three chickens in every ten. The missed birds are, according to Butler, 'scalded alive.' They 'flop, scream, kick, and their eyeballs pop out of their heads.' Often they come out 'with broken bones and disfigured and missing body parts because they've struggled so much in the tank.'[19] When there were mechanical failures, the supervisor would refuse to stop the line, even though he knew that chickens were going into the scalding tank alive or were having their legs broken by malfunctioning equipment.

When people are under pressure and angry with their boss or frustrated about their working conditions, they can do strange things. In January 2003, Butler made a public statement describing workers pulling chickens apart, stomping on them, beating them, running over them on purpose with a fork-lift truck, and even blowing them up with dry ice 'bombs.' Tyson dismissed the statement as the 'outrageous' inventions of a disgruntled worker who had lost his job.

It's true that Butler has a conviction for burglary and has had other problems with the law. But eighteen months after Butler made these supposedly 'outrageous' claims, a videotape secretly filmed at another KFC-supplying slaughterhouse, in Moorfield, West Virginia, made his claims a lot more credible. The slaughterhouse, operated by Pilgrim's Pride, the second largest chicken producer in the nation, had won KFC's 'Supplier of the Year' Award. The tape, taken by an undercover investigator working for People for the Ethical Treatment of Animals, showed slaughterhouse workers behaving in ways quite similar to those described by Butler: slamming live chickens into walls, jumping up and down on them, and drop-kicking them as if they were footballs. The undercover investigator said that, beyond what he had been able to catch on camera, he had witnessed 'hundreds' of acts of cruelty. Workers had ripped off a bird's head to write graffiti in blood, plucked feathers off live chickens to 'make it snow,' suffocated a chicken by tying a latex glove over its head, and squeezed birds like water balloons to spray feces over other birds. The investigator thought that the workers did this because they were bored or needed to vent their frustrations at the nature of the work. Evidently, their work had desensitized them to animal suffering.

The only significant difference between the behavior of the workers at Moorfield and that described by Butler at Grannis was that the behavior at Moorfield was caught on tape. Unable to dismiss the evidence of cruelty, Pilgrim's Pride said that it was 'appalled.'[20] But neither Pilgrim's Pride nor Tyson Foods, the two largest suppliers of chicken in America, have done anything to address the root cause of the problem: unskilled, low-paid workers doing dirty, bloody work, often in stifling heat, under constant pressure to keep the killing lines moving no matter what so that they can slaughter up to 90,000 animals every shift.

NOTES

1. Jennifer Viegas, 'Chickens worry about the future,' *Discovery News*, https://www.abc.net.au/science/articles/2005/07/15/1415178.htm

2. Susan Milius, 'The science of eeeeek: what a squeak can tell researchers about life, society, and all that,' *Science News*, Sept 12, 1998; available at https://www.questia.com/magazine/1G1-21156998/the-science-of-eeeeek-what-a-squeak-can-tell-researchers

3. T. C. Danbury et al., 'Self-selection of the analgesic drug carprofen by lame broiler chickens,' *Veterinary Record*, 146 (11 March 2000), pp. 307–11.

4. Jonathan Leake, 'The Secret Lives of Moody Cows,' *Sunday Times*, February 27, 2005.

5. National Chicken Council, Animal Welfare Guidelines and Audit Checklist, Washington, DC, March 2003, available at https://thepoultrysite.com/articles/animal-welfare-guidelines-and-audit-checklist. On p. 6 it is stated that '. . . density shall not exceed 8.5 pounds live weight per square foot.' Since the average market weight in 2004 was 5 pounds (see www.nationalchickencouncil.com/statistics/stat_detail.cfm?id=2) this is equivalent to 85 square inches per bird.

6. M. O. North and Bell D. D., *Commercial Chicken Production Manual*, 4th edition (New York: Van Nostrand Reinhold, 1990), p. 456.

7. John Vidal, *McLibel: Burger Culture on Trial* (London: Pan Books, 1997), p. 311.

8. See H. L. Brodie et al., 'Structures for Broiler Litter Manure Storage,' Fact Sheet 416, Maryland Cooperative Extension, www.agnr.umd.edu/users/bioreng/fs416.htm, refer, without any suggestion of criticism, to delaying manure cleanout for three years. See also Anon., 'Animal Waste Management Plans,' *Delaware Nutrient Management Notes*, Delaware Department of Agriculture, vol. 1, no. 7 (July 2000), where the calculations are based on 90 percent of the litter remaining in place for two years.

9. C. Berg, 'Foot-Pad Dermatitis in Broilers and Turkeys,' *Veterinaria* 36 (1998); G. J. Wang, C. Ekstrand, and J. Svedberg, 'Wet Litter and Perches as Risk Factors for the Development of Foot Pad Dermatitis in Floor-Housed Hens,' *British Poultry Science* 39 (1998), pp. 191–7; C. M. Wathes, 'Aerial Emissions from Poultry Production,' *World Poultry Science Journal* 54 (1998), pp. 241–51; Kristensen and Wathes, op cit; S. Muirhead, 'Ammonia Control Essential to Maintenance of Poultry Health,' *Feedstuffs* (April 13, 1992), p. 11. On blindness caused by ammonia, see also Michael P. Lacy, 'Litter Quality and Performance,' University of Georgia College of Agriculture and Environmental Sciences, https://thepoultrysite.com/articles/litter-quality-and-broiler-performance and Karen Davis, *Prisoned Chickens, Poisoned Eggs: An Inside Look at the Modern Poultry Industry*, (Summertown, TN: Book Publishing Company, 1996), pp. 62–4, 92, 96–8.

10. G. Havenstein, P. Ferket, and M. Qureshi, 'Growth, livability, and feed conversion of 1957 versus 2001 broilers when fed representative 1957 and 2001 broiler diets,' *Poultry Science* 82 (2003), pp. 1500–1508.

11. S. C. Kestin, T. G. Knowles, A. E. Tinch, and N. G. Gregory, 'Prevalence of Leg Weakness in Broiler Chickens and its Relationship with Genotype,' *Veterinary Record* 131 (1992), pp. 190–9.

12. Quoted in the *Guardian*, October 14, 1991.

13. John Webster, *Animal Welfare: A Cool Eye Towards Eden* (Oxford: Blackwell Science, 1995), p. 156.

14. G. T. Tabler and A. M. Mendenhall, 'Broiler Nutrition, Feed Intake and Grower Economics,' *Avian Advice* 5(4) (Winter 2003), p. 9.

15. J. Mench, 'Broiler breeders: feed restriction and welfare,' *World's Poultry Science Journal*, vol. 58 (2002), pp. 23–9.

16. I. J. H. Duncan, 'The Assessment of Welfare During the Handling and Transport of Broilers,' in J. M. Faure and A. D. Mills (eds.), *Proceedings of the Third European Symposium on Poultry Welfare* (Tours, France: French Branch of the World Poultry Science Association, 1989), pp. 79–91; N. G. Gregory and L. J. Wilkins, 'Skeletal Damage and Bone Defects During Catching and Processing,' in *Bone Biology and Skeletal Disorders in Poultry*. C. C. Whitehead, ed. (Abingdon, England: Carfax Publishing, 1992). Cited from A COK Report: Animal Suffering in the Broiler Industry.

17. Freedom of Information Act #94–363, Poultry Slaughtered, Condemned, and Cadavers, 6/30/94; cited in United

Poultry Concerns, 'Poultry Slaughter: The Need for
Legislation', http://www.upc-online.org/slaughter/
slaughter3web.pdf.

18. 'Tyson to Probe Chicken-slaughter Methods,'
Associated Press, May 25, 2005.

19. Signed statement of Tyson employee, Virgil Butler,
January 30, 2003.

20. Donald G. McNeil Jr, 'KFC Supplier Accused of Animal
Cruelty,' *New York Times*, July 20, 2004.

The Oxford Vegetarians:
A Personal Account

People coming together more or less by accident can have a catalytic effect on each other, so that each achieves more than he or she would have done alone. The Bloomsbury Group – G. E. Moore, Virginia and Leonard Woolf, E. M. Forster, J. M. Keynes, Vanessa and Clive Bell, Lytton Strachey and others – is a famous example. It would be immodest to suggest that the group of vegetarians who were together in Oxford from 1969 to about 1971 can compare with these illustrious figures; yet if the Animal Liberation movement ever succeeds in transforming our attitudes to other species, the Oxford Vegetarians may one day be seen to have been a significant force.

My wife, Renata, and I arrived in Oxford in October 1969. I had come to do a graduate degree in philosophy – the natural climax to the education of an Australian philosophy student preparing for an academic career. My interests were in ethics and political philosophy, but the connection between my philosophical studies and my everyday life would have been hard to discern. My day-to-day existence and my ethical beliefs were much like those of other students. I had no distinctive views about animals, or the ethics of our treatment of them. Like most people, I disapproved of cruelty to animals, but I was not greatly concerned about it. I assumed that the RSPCA and the government could be relied upon to see that cruelty to animals was an isolated occurrence. I thought of vegetarians as, at best, other-worldly idealists, and at worst, cranks. Animal welfare I regarded as a cause for kindly old ladies rather than serious political reformers.

The crack in my complacency about our relations with animals began in 1970 when I accidentally met one of the Oxford group, Richard Keshen, a Canadian, who was also a graduate student in philosophy. He and I were attending lectures given by Jonathan Glover, a Fellow of New College, on free will, determinism, and moral responsibility. They were stimulating lectures, and when they finished a few students often remained behind to ask questions or discuss points with the lecturer. After one particular lecture, Richard and I were among this small group and we left together, discussing the issue further. It was lunchtime, and Richard suggested we go to his college, Balliol, and continue our conversation over lunch. When it came to selecting our meal, I noticed that Richard asked if the spaghetti sauce had meat in it, and when told that it had, took a meatless salad. So when we had talked enough about free will and determinism, I asked Richard why he had avoided meat. That began a discussion that was to change my life.

The change did not take place immediately. What

Richard Keshen told me about the treatment of farm animals, combined with his arguments against our neglect of the interests of animals, gave me a lot to think about, but I was not about to change my diet overnight. Over the next two months Renata and I met Richard's wife, Mary, and two other Canadian philosophy students, Roslind and Stanley Godlovitch, who had been responsible for Richard and Mary becoming vegetarians. Ros and Stan had become vegetarians a year or two earlier, before reaching Oxford. They had come to see our treatment of nonhuman animals as analogous to the brutal exploitation of other races by whites in earlier centuries. This analogy they now urged on us, challenging us to find a morally relevant distinction between humans and nonhumans which could justify the differences we make in our treatment of those who belong to our own species and those who do not.

During these two months, Renata and I read Ruth Harrison's pioneering attack on factory farming, *Animal Machines*. I also read an article which Ros Godlovitch had recently published in the academic journal *Philosophy*. She was in the process of revising it for republication in a book which she, Stan, and John Harris, another vegetarian philosophy student at Oxford, were editing. Ros was a little unsure about the revisions she was making, and I spent a lot of time trying to help her clarify and strengthen her arguments. In the end she went her own way, and I do not think any of my suggestions were incorporated into the revised version of the article as it appeared in *Animals, Men and Morals* – but in the process of putting her arguments in their strongest possible form, I had convinced myself that the logic of the vegetarian position was irrefutable. Renata and I decided that if we were to retain our self-respect and continue to take moral issues seriously, we should cease to eat animals.

Through the Keshens and Godlovitches we got to know other members of a loose group of vegetarians. Several of them lived together in a rambling old house with a huge vegetable garden. Among the residents of this semi-communal establishment were John Harris and two other contributors to *Animals, Men and Morals*, David Wood and Michael Peters. Philosophically we agreed on little but the immorality of our present treatment of animals. David Wood was interested in Continental philosophy, Michael Peters in Marxism and structuralism, Richard Keshen's favorite philosopher was Spinoza, Ros Godlovitch was still developing her basic position – she had not studied philosophy as an undergraduate and only became involved in it as a result of her interest in the ethics of our relations with animals – and Stan Godlovitch refused to work on moral philosophy, restricting himself to the philosophy of biology. I was more in the mainstream of Anglo-American philosophy than any of the others, and in moral philosophy I took a much more utilitarian line than they did.

Also around Oxford at that time were Richard Ryder, Andrew Linzey and Stephen Clark. Richard Ryder was working at the Warneford Hospital, in Oxford. He had written a leaflet on 'Speciesism' – the first use of the term as far as I know – and now was writing an essay on animal experimentation for *Animals, Men and Morals*. Later he developed this work into his splendid attack on animal experimentation, *Victims of Science*. He was also organizing a 'ginger group' within the RSPCA, with the aim of getting that then extremely conservative body to eject its fox-hunters and take a stronger stance on other issues. That seemed a very long shot, then. I was introduced to Richard Ryder through Ros Godlovitch, and from him I learned a lot about animal experimentation. At the time, our positions were the mirror image of each other – I was a vegetarian, but not a strong opponent of animal experimentation, because I naively thought most experiments were necessary to save lives, and therefore justified on utilitarian grounds. Richard Ryder, on the other hand, was not then a vegetarian, but was opposed to animal experimentation because of the extreme suffering it often involved.

Andrew Linzey was interested in the animal issue from the point of view of Christian theology, which was not the concern of most of the group, for we were a non-religious lot. His book, *Animal Rights*, was published by the SCM Press in 1976. Stephen Clark was a Fellow of All Souls College, Oxford, during this period, but I did not get to know him until much later, after he had written *The Moral Status of Animals*, which appeared in 1977.

Animals, Men and Morals, the first of all these books, appeared in 1971. We had great hopes for it, for it demanded a revolutionary change in our attitudes to, and treatment of, nonhuman animals. I think Ros Godlovitch, especially, thought the book might trigger off a widespread protest movement. In the light of these expectations, the book's reception was profoundly disappointing. The major newspapers and weeklies ignored it. In the *Sunday Times*, for example, it was mentioned only in the 'In Brief' column – just one short paragraph of exposition, without a comment. Our ideas seemed to be too radical to be taken seriously by the staid British press.

At the time, the virtual silence which met the British publication of *Animals, Men and Morals* seemed a severe setback. Yet it turned out to be the first of a chain of events that led me to write *Animal Liberation*. Some time after *Animals, Men and Morals* appeared in England, the Godlovitches received some better news: Taplinger had agreed to publish an American edition. But would the book get more attention in America than in Britain? I determined to do my best to see that it would. I had in any case been wanting to write something to make people more aware of the injustice of our treatment of animals, but had been deterred from doing so by the feeling that since so many of my ideas had come from others, and especially from Ros, I should allow her to publish them. Now I thought of a way to satisfy my own desire to do something to make people aware of the issue while at the same time helping to get my friends' ideas the attention they deserved but had not received. I would write a long review article, based on *Animals,*

Men and Morals, but drawing the views of the several contributors together into a single coherent philosophy of Animal Liberation. There was only one place I knew of in America where such a review article might appear: *The New York Review of Books*.

I wrote to the editors of *The New York Review*, describing the book and the review I would write. I did not know what answer to expect, since I had had no previous contact with them, and they would never have heard of me. I knew they were open to novel and radical ideas, but did they perhaps accept contributions only from people they knew? Would the idea of animal liberation seem ridiculous to them?

Robert Silvers' reply was guardedly encouraging. The idea was intriguing, and he would like to see the article, though he could not undertake to publish it. That was all the encouragement I needed, however, and the article was soon written and accepted. Entitled 'Animal Liberation,' it appeared in April 1973. I was soon receiving enthusiastic letters from people who seemed to have been waiting for their feelings about the mistreatment of animals to be given a coherent philosophical backing.

Among the letters was one from a leading New York publisher, who suggested that I develop the ideas sketched in the article into a full-length book. Although my review had helped *Animals, Men and Morals* become better known in America – it eventually went into a paperback edition there, something that never happened in Britain – there was obviously room for a different kind of book, more systematic in its approach than a compilation of articles by different authors can be. There was also a need for factual research to be done on factory farming and experimentation in America, since the data in both *Animal Machines* and *Animals, Men and Morals* was largely British. By this time I knew that I would soon be leaving Oxford, for I had accepted a visiting position at New York University, which would make a good base for this kind of research. So during our last summer in Oxford, I began work on *Animal Liberation*.

The Oxford Vegetarians had already begun to scatter. Most of the students had finished their degrees. John Harris had gone to Manchester, David Wood to Warwick, Richard and Mary Keshen returned to Canada, and Stan and Ros Godlovitch had separated, Stan to return to Canada while Ros remained in Oxford. We had built strong bonds of friendship and affection, based in part on our respect for each other's ethical commitment to vegetarianism. Along with our ideas about animals we had shared an enjoyment of nature, often walking together by the Thames and through the Oxfordshire countryside. On walks with Stan I learned a little about birds, and from both Stan and Richard I learned to grow a few of my own vegetables. We had shared many meals, and our recipes as well, for as vegetarian cooks we all still had many things to learn.

It is too early to say what influence the group has had. If the books we produced have helped change the animal welfare movement, then our influence has been important. But it is difficult to single out causes for events as broad and disparate as the revitalization of the animal welfare movement. The broader ecology movement of the late sixties and early seventies obviously had a lot to do with it and there were many others, not connected with Oxford, who worked long and hard for this revitalization. Whatever the historian's verdict on the influence of a group of young vegetarians at Oxford in the early seventies, however, I know that had the Keshens and Godlovitches not been in Oxford when I was there, I would have missed an episode of my life that has put its mark on almost everything I have thought and written – let alone everything I have cooked and eaten – ever since.

A Vegetarian Philosophy

Issues regarding eating meat were highlighted in 1997 by the longest trial in British legal history. McDonald's Corporation and McDonald's Restaurants Limited v. Steel and Morris, better known as the 'McLibel' trial, ran for 515 days and heard 180 witnesses. In suing Helen Steel and David Morris, two activists involved with the London Greenpeace organization, McDonald's put on trial the way in which its fast-food products are produced, packaged, advertised, and sold, as well as their nutritional value, the environmental impact of producing them, and the treatment of the animals whose flesh and eggs are made into that food.

The case provided a remarkable opportunity for weighing up evidence for and against modern agribusiness methods. The leaflet 'What's Wrong with McDonald's' that provoked the defamation suit had a row of McDonald's arches along the top of each page. Two of these arches bore the words 'McMurder' and 'McTorture.' One section below was headed 'In what way are McDonald's responsible for torture and murder?' The leaflet answered the question as follows:

The menu at McDonald's is based on meat. They sell millions of burgers every day in 55 countries throughout the world. This means the constant slaughter, day by day, of animals born and bred solely to be turned into McDonald's products. Some of them – especially chickens and pigs – spend their lives in the entirely artificial conditions of huge factory farms, with no access to air or sunshine and no freedom of movement. Their deaths are bloody and barbaric.

McDonald's claimed that the leaflet meant that the company was responsible for the inhumane torture and murder of cattle, chicken, and pigs, and that this was defamatory. In considering this claim, Mr Justice Bell based his judgment on what he took to be attitudes that were generally accepted in Britain. Thus for the epithet 'McTorture' to be justified, he held, it would not be enough for Steel and Morris to show that animals were under stress or suffered some pain or discomfort:

> Merely containing, handling and transporting an animal may cause it stress; and taking it to slaughter certainly may do so. But I do not believe that the ordinary reasonable person believes any of these things to be cruel, provided that the necessary stress, or discomfort or even pain is kept to a reasonably acceptable level. That ordinary person may know little about the detail of farming and slaughtering methods

but he must find a certain amount of stress, discomfort or even pain acceptable and not to be criticised as cruel.

By the end of the trial, however, Mr Justice Bell found that the stress, discomfort, and pain inflicted on some animals amounted to more than this acceptable level, and hence did constitute a 'cruel practice' for which McDonald's was 'culpably responsible.' Chickens, laying hens and sows kept in individual stalls, he said, suffered from 'severe restriction of movement' which 'is cruel.' He also found a number of other cruel practices in the production of chickens, including the restricted diet fed to breeding birds, which leaves them permanently hungry; the injuries inflicted on chickens by catchers stuffing 600 birds an hour into crates to take them to slaughter; and the failure of the stunning apparatus to ensure that all birds are stunned before they have their throats cut. Judging by entirely conventional moral standards, Mr Justice Bell held these practices to be cruel, and McDonald's to be culpably responsible for them.

It was not libelous to describe McDonald's as 'McTorture,' because the charge was substantially true. What follows from this judgment about the morality of buying and eating intensively raised chickens, pig products that come from the offspring of sows kept in stalls, or eggs laid by hens kept in battery cages? Surely that, too, must be wrong?

This claim has been challenged. At a conference dinner some years ago I found myself sitting opposite a

Buddhist philosopher from Thailand. As we helped ourselves to the lavish buffet, I avoided the various forms
of meat being offered, but the Thai philosopher did not.
When I asked him how he reconciled the dinner he had
chosen with the first precept of Buddhism, which tells
us to avoid harming sentient beings, he told me that in
the Buddhist tradition it is wrong to eat meat only if you
have reason to believe that the animal was killed specially for you. The meat he had taken, however, was not
from animals killed specially for him; the animals
would have died anyway, even if he were a strict vegetarian or had not been in that city at all. Hence, by
eating it, he was not harming any animals.

I was unable to convince my dinner companion that
this defense of meat-eating was better suited to a time
when a peasant family might kill an animal especially to
have something to put in the begging bowl of a wandering
monk than it is to our own era. The flaw in the defense is
the disregard of the link between the meat I eat today and
the future killing of animals. Granted, the chicken lying
in the supermarket freezer today would have died even if
I had never existed; but the fact that I take the chicken
from the freezer, and ignore the tofu on a nearby shelf,
has something to do with the number of chickens, or
blocks of tofu, the supermarket will order next week and
thus contributes, in a small way, to the future growth or
decline of the chicken and tofu industries. That is what
the laws of supply and demand are all about.

Some defenders of a variant of the ancient Buddhist
line may still want to argue that one chicken fewer sold

makes no perceptible difference to the chicken producers, and therefore there can be nothing wrong with
buying chicken. The division of moral responsibility in
a situation of this kind does raise some interesting
issues, but it is a fallacy to argue that a person can do
wrong only by making a perceptible harm. The Oxford
philosopher Jonathan Glover has explored the implica-
tions of this refusal to accept the divisibility of
responsibility in an entertaining article called 'It makes
no difference whether or not I do it' (*Proceedings of the
Aristotelian Society*, 1975).

Glover imagines that in a village, 100 people are
about to eat lunch. Each has a bowl containing 100
beans. Suddenly, 100 hungry bandits swoop down on
the village. Each bandit takes the contents of the bowl
of one villager, eats it, and gallops off. Next week, the
bandits plan to do it again, but one of their number is
afflicted by doubts about whether it is right to steal from
the poor. These doubts are set to rest by another of their
number who proposes that each bandit, instead of eating the entire contents of the bowl of one villager,
should take one bean from every villager's bowl. Since
the loss of one bean cannot make a perceptible difference to any villager, no bandit will have harmed anyone.
The bandits follow this plan, each taking a solitary bean
from 100 bowls. The villagers are just as hungry as they
were the previous week, but the bandits can all sleep
well on their full stomachs, knowing that none of them
has harmed anyone.

Glover's example shows the absurdity of denying

that we are each responsible for a share of the harms we collectively cause, even if each of us makes no perceptible difference. McDonald's has a far bigger impact on the practices of the chicken, egg, and pig industries than any individual consumer; but McDonald's itself would be powerless if no one ate at its restaurants. Collectively, all consumers of animal products are responsible for the existence of the cruel practices involved in producing them. In the absence of special circumstances, a portion of this responsibility must be attributed to each purchaser.

Without in any way departing from a conventional moral attitude toward animals, then, we have reached the conclusion that eating intensively produced chicken, battery eggs, and some pig products is wrong. This is, of course, well short of an argument for vegetarianism. Mr Justice Bell found 'cruel practices' only in these areas of McDonald's food production. But he did not find that McDonald's beef is 'cruelty-free.' He did not consider that question, because he drew a distinction between McDonald's responsibility for practices in the beef and dairy industries and those in the chicken, egg, and pig industries. McDonald's chickens, eggs, and pig products are supplied by a relatively small number of very large producers, over whose practices the corporation could quite easily have a major influence. On the other hand, McDonald's beef and dairy requirements came from a very large number of producers; and in respect of whose methods, Mr Justice Bell held, 'there was no evidence from which I could infer that [McDonald's] would have any effective influence, should it try to exert it.' Whatever one may think of that view – it seems highly implausible to me – the judge, in accepting it, decided not to address the evidence presented to him of cruelty in the raising of cattle, so that no conclusions either way can be drawn.

This does not mean that the trial itself had nothing to say about animal suffering in general. McDonald's called as a witness Mr David Walker, chief executive of one of McDonald's major United Kingdom suppliers, McKey Food Services Ltd. In cross-examination, Helen Steel asked Walker whether it was true that, 'as the result of the meat industry, the suffering of animals is inevitable.' Walker replied: 'The answer to that must be "yes." '

Walker's admission raises a serious question about the ethics of the meat industry: how much suffering are we justified in inflicting on animals in order to turn them into meat, or to use their eggs or milk?

The case for vegetarianism is at its strongest when we see it as a moral protest against our use of animals as mere things, to be exploited for our convenience in whatever way makes them most cheaply available to us. Only the tiniest fraction of the tens of billions of farm animals slaughtered for food each year – the figure for the United States alone is nine billion – were treated during their lives in ways that respected their interests. Questions about the wrongness of killing in itself are not relevant to the moral issue of eating meat or eggs from factory-farmed animals, as most people in developed countries do. Even when animals are roaming freely over large areas, as sheep and cattle do in Australia, operations like hot-iron branding, castration, and dehorning are carried out without any regard for the animals' capacity to suffer. The same is true of handling and transport prior to slaughter. In the light of these facts, the issue to focus on is not whether there are some circumstances in which it could be right to eat meat, but on what we can do to avoid contributing to this immense amount of animal suffering.

The answer is to boycott all meat and eggs produced by large-scale commercial methods of animal production, and encourage others to do the same. Consideration for the interests of animals alone is enough justification for this response, but the case is further strengthened by the environmental problems that the meat industry causes. Although Mr Justice Bell found that the allegations directed at McDonald's regarding its contribution to the destruction of rain forests were not true, the meat industry as a whole can take little comfort from that, because Bell accepted evidence that cattle-ranching, particularly in Brazil, had contributed to the clearing of vast areas of rain forest. The problem for David Morris and Helen Steel was that they did not convince the judge that the meat used by McDonald's came from these regions. So the meat industry as a whole remains culpable for the loss of rain forest and for all the consequences of that, from global warming to the deaths of indigenous people fighting to defend their way of life.

Environmentalists are increasingly recognizing that the choice of what we eat is an environmental issue.

Animals raised in sheds or on feedlots eat grains or soybeans, and they use most of the food value of these products simply in order to maintain basic functions and develop unpalatable parts of the body like bones and skin. To convert eight or nine kilos of grain protein into a single kilo of animal protein wastes land, energy, and water. On a crowded planet with a growing human population, that is a luxury that we are becoming increasingly unable to afford.

Intensive animal production is a heavy user of fossil fuels and a major source of pollution of both air and water. It releases large quantities of methane and other greenhouse gases into the atmosphere. We are risking unpredictable changes to the climate of our planet – which means, ultimately, the lives of billions of people, not to mention the extinction of untold thousands of species of plants and animals unable to cope with changing conditions – for the sake of more hamburgers. A diet heavy in animal products, catered to by intensive animal production, is a disaster for animals, the environment, and the health of those who eat it.

A RECIPE

This recipe is vegan, very simple, nutritious, and tasty. It's also eaten by hundreds of millions of people every day.

DAL

- 2 tablespoons oil
- 1 onion, chopped

- 2 cloves garlic, crushed
- 1 cup dry red lentils
- 3 cups water
- bay leaf
- 1 cinnamon stick
- 1 teaspoon medium curry powder or to taste
- 1 14-ounce can of chopped tomatoes or equivalent chopped fresh tomatoes
- 2 ounces creamed coconut or half cup coconut milk (optional)
- Juice of lemon (optional)
- Salt to taste

In a deep frying pan, heat the oil and fry the onion and garlic until translucent. Add the lentils and fry them for a minute or two, then add the water, bay leaf, cinnamon stick, and curry powder. Stir, bring to a boil, then let simmer for twenty minutes, adding a little more water from time to time if it gets dry. Add the tomatoes and simmer another ten minutes. By now the lentils should be very soft. Add the creamed coconut or coconut milk and lemon juice, if using, and salt to taste. Remove cinnamon stick and bay leaf before serving.

The final product should flow freely – add more water if it is too thick. It is usually served over rice, with some lime pickle and mango chutney. Sliced banana is another good accompaniment, and so too are pappadams.

If Fish Could Scream

When I was a child, my father used to take me for walks, often along a river or by the sea. We would pass people fishing, perhaps reeling in their lines with struggling fish hooked at the end of them. Once I saw a man take a small fish out of a bucket and impale it, still wriggling, on an empty hook to use as bait.

Another time, when our path took us by a tranquil stream, I saw a man sitting and watching his line, seemingly at peace with the world, while next to him, fish he had already caught were flapping helplessly and gasping in the air. My father told me that he could not understand how anyone could enjoy an afternoon spent taking fish out of the water and letting them die slowly.

These childhood memories flooded back when I read *Worse things happen at sea: the welfare of wild-caught fish*, a breakthrough report released last month on fishcount.org.uk. In most of the world, it is accepted that if animals are to be killed for food, they should be killed without suffering. Regulations for slaughter generally require that animals be rendered instantly unconscious before they are killed, or death should be brought about instantaneously, or, in the case of ritual slaughter, as close to instantaneously as the religious doctrine allows.

Not for fish. There is no humane slaughter requirement for wild fish caught and killed at sea, nor, in most places, for farmed fish. Fish caught in nets by trawlers are dumped on board the ship and allowed to suffocate. Impaling live bait on hooks is a common commercial practice: long-line fishing, for example, uses hundreds or even thousands of hooks on a single line that may be 50–100 kilometers long. When fish take the bait, they are likely to remain caught for many hours before the line is hauled in.

Likewise, commercial fishing frequently depends on gill nets – walls of fine netting in which fish become snared, often by the gills. They may suffocate in the net, because, with their gills constricted, they cannot breathe. If not, they may remain trapped for many hours before the nets are pulled in.

The most startling revelation in the report, however, is the staggering number of fish on which humans inflict these deaths. By using the reported tonnages of the various species of fish caught, and dividing by the estimated average weight for each species, Alison Mood, the report's author, has put together what may well be the first-ever systematic estimate of the size of the annual global capture of wild fish. It is, she calculates, in the order of *one trillion*, although it could be as high as 2.7 trillion.

To put this in perspective, the United Nations Food and Agriculture Organization estimates that 60 billion animals are killed each year for human consumption – the equivalent of about nine animals for each human being on the planet. If we take Mood's lower estimate of one trillion, the comparable figure for fish is 150. This does not include billions of fish caught illegally nor unwanted fish accidentally caught and discarded, nor does it count fish impaled on hooks as bait.

Many of these fish are consumed indirectly – ground up and fed to factory-farmed chicken or fish. A typical salmon farm churns through 3–4 kilograms of wild fish for every kilogram of salmon that it produces.

Let's assume that all this fishing is sustainable, though of course it is not. It would then be reassuring to believe that killing on such a vast scale does not matter, because fish do not feel pain. But the nervous systems of fish are sufficiently similar to those of birds and mammals to suggest that they do. When fish experience something that would cause other animals physical pain, they behave in ways suggestive of pain, and the change in behavior may last several hours. (It is a myth that fish have short memories.) Fish learn to avoid unpleasant experiences, like electric shocks. And pain-killers reduce the symptoms of pain that they would otherwise show.

Victoria Braithwaite, a professor of fisheries and biology at Pennsylvania State University, has probably spent more time investigating this issue than any other scientist. Her recent book *Do Fish Feel Pain?* shows that fish are not only capable of feeling pain, but also are a lot smarter than most people believe. Last year, a scientific panel advising the European Union concluded that the preponderance of the evidence indicates that fish do feel pain.

Why are fish the forgotten victims on our plate? Is it because they are cold-blooded and covered in scales? Is it because they cannot give voice to their pain? Whatever the explanation, the evidence is now accumulating that commercial fishing inflicts an unimaginable amount of pain and suffering. We need to learn how to capture and kill wild fish humanely – or, if that is not possible, to find less cruel and more sustainable alternatives to eating them.

The Case for Going Vegan

Can we defend the things we do to animals? Christians, Jews and Moslems may appeal to scripture to justify their dominion over animals. Once we move beyond a religious outlook, we have to face 'the animal question' without any prior assumption that animals were created for our benefit or that our use of them has divine sanction. If we are just one species among others that have evolved on this planet, and if the other species include billions of nonhuman animals who can also suffer, or conversely can enjoy their lives, should our interests always count for more than theirs?

Of all the ways in which we affect animals, the one most in need of justification today is raising them for food. Far more animals are affected by this than by any other human activity. In the United States alone, the number of animals raised and killed for food every year is now nearly ten billion.[1] All of this is, strictly speaking, unnecessary. In developed countries, where we have a wide choice of foods, no one needs to eat meat. Many

1 Surprisingly, the number of farm animals killed in the US peaked around the time this article was written, and has subsequently fallen to 9.1 billion.

studies show that we can live as healthily, or more
healthily, without it. We can also live well on a vegan
diet, consuming no animal products at all. (Vitamin B12
is the only essential nutrient not available from plant
foods, and it is easy to take a supplement obtained from
vegan sources.)

Ask people what the main ethical problem about eating animals is, and most will refer to killing. That is an
issue, of course, but at least as far as modern industrial
animal productions is concerned, there is a more
straightforward objection. Even if there were nothing
wrong with killing animals because we like the taste of
their flesh, we would still be supporting a system of agriculture that inflicts prolonged suffering on animals.

Chickens raised for meat are kept in sheds that hold
more than 20,000 birds. The level of ammonia in the air
from their accumulated droppings stings the eye and
hurts the lungs. Slaughtered at only 45 days old, their
immature bones can hardly bear the weight of their
bodies. Some collapse and, unable to reach food or
water, soon die, their fate irrelevant to the economics of
the enterprise as a whole. Catching, transport and
slaughter are brutal processes in which the economic
incentives all favor speed, and the welfare of the birds
plays no role at all.

Laying hens are crammed into wire cages so small
that even if there were just one per cage, she would be
unable to stretch her wings. But there are usually at
least four hens per cage, and often more. Under such
crowded conditions, the more aggressive birds peck at

the weaker hens in the cage, who are unable to escape. To prevent this pecking leading to fatalities, producers sear off all the birds' beaks with a hot blade. A hen's beak is full of nerve tissue – it is her principal means of relating to her environment – but no anesthetic or analgesic is used to relieve the pain.

Pigs may be the most intelligent and sensitive of the animals we commonly eat. In today's factory farms, pregnant sows are kept in crates so narrow that they cannot turn around, or even walk more than a step forward or backward. They lie on bare concrete without straw or any other form of bedding. They have no way of satisfying their instinct to build a nest just before giving birth. The piglets are taken from the sow as soon as possible, so that she can be made pregnant again, but they too are kept indoors, on bare concrete, until they are taken to slaughter.

Beef cattle spend the last six months of their lives in feedlots, on bare dirt, eating grain that is not suitable for their digestion, fed steroids to make them put on more muscle, and antibiotics to keep them alive. They have no shade from the blazing summer sun, or shelter from winter blizzards.

But what, you may ask, is wrong with milk and other dairy products? Don't the cows have a good life, grazing on the fields? And we don't have to kill them to get milk. But most dairy cows are now kept inside, and do not have access to pasture. Like human females, they do not give milk unless they have recently had a baby, and so dairy cows are made pregnant every year. The calf is taken away from its mother just hours after birth, so that it will not drink the milk intended for humans. If it is male, it may be killed immediately, or raised for veal, or perhaps for hamburger beef. The bond between a cow and her calf is strong, and she will often call for the calf for several days after it is taken away.

★

In addition to the ethical question of our treatment of animals, there is now a powerful new argument for a vegan diet. Ever since Frances Moore Lappé published *Diet for a Small Planet* in 1971, we have known that modern industrial animal production is extremely wasteful. Pig farms use six pounds of grain for every pound of boneless meat they produce. For beef cattle in feedlots, the ratio is 13:1. Even for chickens, the least inefficient factory-farmed meat, the ratio is 3:1.

Lappé was concerned about the waste of food and the extra pressure on arable land this involves, since we could be eating the grain and soybeans directly, and feeding ourselves just as well from much less land. Now global warming sharpens the problem. Most Americans think that the best thing they could do to cut their personal contribution to global warming would be to swap their family car for a fuel-efficient hybrid like the Toyota Prius. Gidon Eshel and Pamela Martin, researchers at the University of Chicago, have calculated that while this would indeed lead to a reduction in emissions of about 1 ton of carbon dioxide per driver, switching from the typical U.S. diet to a vegan diet would save the equivalent of almost 1.5 tons of carbon dioxide per person. Vegans are therefore doing significantly less damage to our climate than those who eat animal products.[2]

*

Is there an ethical way of eating animal products? It is possible to obtain meat, eggs, and dairy products from animals who have been treated less cruelly, and allowed to eat grass rather than grain or soy. Limiting one's consumption of animal products to these sources also avoids some of the greenhouse gas emissions, although cows kept on grass still emit substantial amounts of methane, a particularly potent contributor to global warming. So *if* there is no serious ethical objection to killing animals, as long as they have had good lives, then being selective about the animal products you eat could provide an ethically defensible diet. It needs care, however. 'Organic,' for instance, says little about animal welfare and hens not kept in cages may still be crowded into a large shed. Going vegan is a simpler choice that sets a clearcut example for others to follow.

2 Gidon Eshel and Pamela Martin, 'Diet, Energy and Global Warming', *Earth Interactions*, 10–009 (2006).

Can Cultured Meat Save the Planet?

In September, California governor Jerry Brown signed a bill mandating that by 2045, all of California's electricity will come from clean power sources. Technological breakthroughs in producing electricity from sun and wind, as well as lowering the cost of battery storage, have played a major role in persuading Californian legislators that this goal is realistic. James Robo, the CEO of the Fortune 200 company NextEra Energy, has predicted that by the early 2020s, electricity from solar farms and giant wind turbines will be cheaper than the operating costs of coal-fired power plants, even when the cost of storage is included.

Can we therefore all breathe a sigh of relief, because technology will save us from catastrophic climate change? Not yet. Even if the world were to move to an entirely clean power supply, and use that clean power to charge up an all-electric fleet of cars, buses and trucks, one major source of greenhouse gas emissions would continue to grow: meat.

The livestock industry now accounts for about 15% of global greenhouse gas emissions, roughly the same as the emissions from the tailpipes of all the world's vehicles. But whereas vehicle emissions can be expected to decline as hybrids and electric vehicles proliferate, global meat consumption is forecast to be 76% greater in 2050 than it has been in recent years. Most of that growth will come from Asia, especially China, where increasing prosperity has led to an increasing demand for meat.

Changing Climate, Changing Diets, a report from the London-based Royal Institute of International Affairs, indicates the threat posed by meat production. At the UN climate change conference held in Cancun in 2010, the participating countries agreed that to allow global temperatures to rise more than 2°C above pre-industrial levels would be to run an unacceptable risk of catastrophe. Beyond that limit, feedback loops will take effect, causing still more warming. For example, the thawing Siberian permafrost will release large quantities of methane, causing yet more warming and releasing yet more methane. Methane is a greenhouse gas that, ton for ton, warms the planet 30 times as much as carbon dioxide.

The quantity of greenhouse gases we can put into the atmosphere between now and mid-century without heating up the planet beyond 2°C – known as the 'carbon budget' – is shrinking steadily. The growing demand for meat means, however, that emissions from the livestock industry will continue to rise, and will absorb an increasing share of this remaining carbon budget. This will, according to *Changing Climate, Changing Diets*, make it 'extremely difficult' to limit the temperature rise to 2°C.

One reason why eating meat produces more greenhouse gases than getting the same food value from plants is that we use fossil fuels to grow grains and soybeans and feed them to animals. The animals use most of the energy in the plant food for themselves, moving, breathing and keeping their bodies warm. That leaves only a small fraction for us to eat, and so we have to grow several times the quantity of grains and soybeans that we would need if we ate plant foods ourselves. The other important factor is the methane produced by ruminants – mainly cattle and sheep – as part of their digestive process. Surprisingly, that makes grass-fed beef even worse for our climate than beef from animals fattened in a feedlot. Cattle fed on grass put on weight more slowly than cattle fed on corn and soybeans, and therefore burp and fart more methane per kilogram of flesh they produce.

If technology can give us clean power, can it also give us clean meat? That term is already in use, by advocates of growing meat at the cellular level. They use it, not to make the parallel with clean energy, but to emphasize that meat from live animals is dirty, because live animals shit. Bacteria from the animals' guts and shit often contaminates the meat. With meat cultured from cells grown in a bioreactor, there is no live animal, no shit, and no bacteria from a digestive system to get mixed into the meat. There is also no methane. Nor is there a living animal to keep warm, move around, or grow body parts that we do not eat. Hence producing meat in this way would be much more efficient, and much cleaner, in the environmental sense, than producing meat from animals.

There are now many startups working on bringing clean meat to market. Plant-based products that have the texture and taste of meat, like the 'Impossible Burger' and the 'Beyond Burger,' are already available in restaurants and supermarkets. Clean hamburger meat, fish, dairy, and other animal products are all being produced without raising and slaughtering a living animal. The price is not yet competitive with animal products, but it is coming down rapidly. Just this week, leading officials from the Food and Drug Administration and the U.S. Department of Agriculture have been meeting to discuss how to regulate the expected production and sale of meat produced by this method

When Kodak, which once dominated the sale and processing of photographic film, decided to treat digital photography as a threat rather than an opportunity, it signed its own death warrant. Tyson Foods and Cargill, two of the world's biggest meat producers, are not making the same mistake. They are investing in companies seeking to produce meat without raising animals. Justin Whitmore, Tyson's executive vice-president, said, 'We don't want to be disrupted. We want to be part of the disruption.'

That's a brave stance for a company that has made its fortune from raising and killing tens of billions of animals, but it is also an acknowledgement that when new technologies create products that people want, they cannot be resisted. Richard Branson, who has invested in the biotech company Memphis Meats, has suggested that in 30 years, we will look back on the present era and be shocked that we killed animals en masse for food. If that happens, technology will have made possible the greatest ethical step forward in the history of our species, saving the planet and eliminating the vast quantity of suffering that industrial farming is now inflicting on animals.

The Two Dark Sides of COVID-19

WITH PAOLA CAVALIERI

The apocalyptic images of the locked-down Chinese city of Wuhan have reached us all. The world is holding its breath over the spread of the new coronavirus, COVID-19, and governments are taking or preparing drastic measures that will necessarily sacrifice individual rights and freedoms for the general good.

Some focus their anger on China's initial lack of transparency about the outbreak. The philosopher Slavoj Žižek has spoken of 'the racist paranoia' at work in the obsession with COVID-19 when there are many worse infectious diseases from which thousands die every day. Those prone to conspiracy theories believe that the virus is a biological weapon aimed at China's economy. Few mention, let alone confront, the underlying cause of the epidemic.

Both the 2003 SARS (Severe Acute Respiratory Syndrome) epidemic and the current one can be traced to China's 'wet markets' – open-air markets where animals are bought live and then slaughtered on the spot for the customers. Until late December 2019, everyone affected by the virus had some link to Wuhan's Huanan Market.

At China's wet markets, many different animals are sold and killed to be eaten: wolf cubs, snakes, turtles, guinea pigs, rats, otters, badgers, and civets. Similar markets exist in many Asian countries, including Japan, Vietnam, and the Philippines.

In tropical and subtropical areas of the planet, wet markets sell live mammals, poultry, fish, and reptiles, crammed together and sharing their breath, their blood and their excrement. As US National Public Radio journalist Jason Beaubien recently reported: 'Live fish in open tubs splash water all over the floor. The countertops of the stalls are red with blood as fish are gutted and filleted right in front of the customers' eyes. Live turtles and crustaceans climb over each other in boxes. Melting ice adds to the slush on the floor. There's lots of water, blood, fish scales, and chicken guts.' Wet markets, indeed.

Scientists tell us that keeping different animals in close, prolonged proximity with one another and with people creates an unhealthy environment that is the probable source of the mutation that enabled COVID-19 to infect humans. More precisely, in such an environment, a coronavirus long present in some animals underwent rapid mutation as it changed from nonhuman host to nonhuman host, and ultimately gained the ability to bind to human cell receptors, thus adapting to the human host.

This evidence prompted China, on 26 January, to impose a temporary ban on wildlife animal trade. It is not the first time that such a measure has been introduced in response to an epidemic. Following the SARS out-break China prohibited the breeding, transport and sale of civets and other wild animals, but the ban was lifted six months later.

Today, many voices are calling for a permanent shutdown of 'wildlife markets.' Zhou Jinfeng, head of China's Biodiversity Conservation and Green Development Foundation, has urged that 'illegal wildlife trafficking' be banned indefinitely and has indicated that the National People's Congress is discussing a bill to outlaw trade in protected species. Focusing on protected species, however, is a ploy to divert public attention away from the appalling circumstances in which animals in wet markets are forced to live and die. What the world really needs is a permanent ban on wet markets.

For the animals, wet markets are hell on earth. Thousands of sentient, palpitating beings endure hours of suffering and anguish before being brutally butchered. This is just one small part of the suffering that humans systematically inflict on animals in every country – in factory farms, laboratories, and the entertainment industry.

If we stop to reflect on what we are doing – and mostly we do not – we are prone to justify it by appealing to the alleged superiority of our species, in much the same way that white people used to appeal to the alleged superiority of their race to justify their subjection of 'inferior' humans. But at this moment, when vital human interests so clearly run parallel to the interests of nonhuman animals, this small part of the suffering we inflict on animals offers us the opportunity for a change of attitudes toward members of nonhuman species.

To achieve a ban on wet markets, we will have to overcome some specific cultural preferences, as well as resistance linked to the fact that a ban would cause economic hardship to those who make their living from the markets. But, even without giving nonhuman animals the moral consideration they deserve, these localized concerns are decisively outweighed by the calamitous impact that ever more frequent global epidemics (and perhaps pandemics) will have.

Martin Williams, a Hong Kong-based writer specializing in conservation and the environment, puts it well: 'As long as such markets exist, the likelihood of other new diseases emerging will remain. Surely, it is time for China to close down these markets. In one fell swoop, it would be making progress on animal rights and nature conservation, while reducing the risk of a "made in China" disease harming people worldwide.'

But we would go further. Historically, tragedies have sometimes led to important changes. Markets at which live animals are sold and slaughtered should be banned not only in China, but all over the world.